"Danger to our country is to be apprehended not so much from the influence of new things as from our forgetting the values of old things." *Henry Ford*

Henry Ford Museum & Greenfield Village provides unique educational experiences based on authentic objects, stories and lives from America's traditions of ingenuity, resourcefulness and innovation. Our purpose is to inspire people to learn from these traditions to help shape a better future.

©1999, Henry Ford Museum & Greenfield Village
All rights reserved.

Editors
Wes Hardin, William S. Pretzer and Susan M. Steele

Text
Nancy E. V. Bryk, Wes Hardin, Andrew K. Johnson, Jeanine Head Miller, Henry Prebys and William S. Pretzer

Design
Savitski Design, Ann Arbor, Michigan

Photo Credits
All photographs are from Henry Ford Museum & Greenfield Village, except where noted.

Printed by University Lithoprinters, Inc., Ann Arbor, Michigan

10 9 8 7 6 5 4 3 2 1

ISBN 0-933728-03-4

Henry Ford Museum & Greenfield Village is an independent, nonprofit educational organization, not affiliated with the Ford Motor Company or the Ford Foundation.

The authors would like to give special thanks to the following people, without whose help this book would not have been possible: Terri Anderson, Ford R. Bryan, Bob Casey, Toby Hines, Margaret Hoover, Terry Hoover, Leo Landis, Minoo Larson, Cathy Latandresse, Christian Overland, Cynthia Read Miller, Romie Minor, Rudy Ruzicska, Julia Saylor-Plumhoff, Linda Skolarus, Alene Soloway and John Wright.

Inside front cover
Henry Ford, with Clara to the right and Edsel to the left, are surrounded by students of the Edison Institute Schools around 1942.

Inside back cover
The first class of the Henry Ford Academy entered the school at the museum in 1997.

Henry Ford Museum &
Greenfield Village
20900 Oakwood Boulevard
Dearborn, MI 48124
313 271-1620

www.hfmgv.org

The Story of Henry Ford Museum & Greenfield Village®

William Clay Ford is shown behind the wheel of his grandfather's famous race car, the *999*, in 1953.

In gratitude for his years of
leadership and support
This book is dedicated to
William Clay Ford

Who joined the Board of Trustees
of The Edison Institute in 1951

Served as Chairman from
1968 to 1989

And has continued to serve as
Chairman Emeritus since 1989

And for whom in 1999 the
Great Hall of Henry Ford Museum
was renamed "The William Clay
Ford Hall of American Innovation"
by the Board of Trustees

vii	**INTRODUCTION**
	STEVEN K. HAMP
1	**THE FOUNDATION**
13	**THE CREATION**
25	**THE DEDICATION**
35	**THE EXPANSION**
53	**THE TRANSITION**
69	**THE TRANSFORMATION**
83	**THE REINVENTION**

INTRODUCTION

In one of his ever-present "jot books," Henry Ford copied a favorite quote from Ralph Waldo Emerson: "Every institution is the lengthened shadow of one man." Perhaps it caught his eye because it satisfactorily represented for him a way to define his own career of institution-building in industry and education. Henry Ford Museum & Greenfield Village, one of the most powerful and enduring institutional creations of Henry Ford's life, still rests under his lengthened shadow—a presence that animates and envelopes the museum's spirit, shape, and contents.

But so, too, do the shadows of many, many others, people contemporary with the founder as well as those who followed in succeeding generations of leadership. Their work extended, but also changed, the institution Henry Ford created. Today, that shadow is a patchwork quilt of contributions and influences, emphases and eras, made up of the additive work of numerous builders and shapers throughout the seven-plus decades of the museum's existence. This volume tells the story of Henry Ford Museum & Greenfield Village.

Poised on the threshold of the new millennium, Henry Ford Museum & Greenfield Village stands as a mature, but not finished, institution. Its founding generation feels antique to us today, belonging to another place and time. Yet, it really was not so long ago when Henry Ford paced the grounds, instructing workers to move his newly acquired historic structures about the landscape until Greenfield Village began to take satisfactory shape. Or when Thomas Edison re-created his most famous invention, the incandescent light bulb, in front of a who's who of pre-Depression Era Americans in the Menlo Park laboratory, on soil freshly relocated from New Jersey to Dearborn, Michigan. Or when George Washington Carver experimented with soybeans, ragweed,

Henry Ford (left) and Thomas Edison at the Menlo Park Laboratory compound in Greenfield Village in 1930. Edison holds a replica of his 1879 light bulb while Ford compares it to a modern light bulb from 1930.

and other botanical species on the museum's grounds, seeking industrial applications from humble plants. Or when Orville Wright planned, with his friend Henry Ford, just where the bicycle shop should stand in which he and his brother, Wilbur, designed and built the first successful powered aircraft. Or when the entirety of the site, both the museum and the village, echoed daily with the sounds of schoolchildren, simultaneously at work and at play, students of the Edison Institute School.

Distant as it might seem today, we are, perhaps, closer in many ways to an understanding of that era than we are of the one we are entering. That's probably because there are several consistent themes that run through the museum's history. The superb collections—the basic resource that fueled Henry's vision—are unique in the world in their depth, breadth, and cultural value, impossible to reassemble today, a three-dimensional record of human achievement. Pioneer public programs, now quite common fare in many museums, were invented here, especially the extravagant weekend special events, the granddaddy of which is the Old Car Festival, nearly 50 years old. Innovative, even radical, educational ventures were born here, the most interesting being the elementary and high school programs that placed children at the very center of the institution. Perhaps most important of all, Henry Ford Museum & Greenfield Village was recognized years ago as a warm and friendly place with a tradition of hospitality that it continues to enjoy, a trademark trait of this museum that has welcomed over 65 million visitors through its doors. Hospitable and educational, children- and family-focused, full of accessible history and memorable events, presented in a unique environment of landmark buildings, priceless collections, and beautiful grounds—these have become signature features of Henry Ford Museum & Greenfield Village.

But, of course, this institution, like any other, must continue to respond to a changing world. The current generation of museum leaders will both maintain and alter the institution—and then give way to the next. Today, our efforts are centered on renewing and expanding the physical plant, increasing the quality of our public presentations, establishing exciting and valuable educational programs and outreach efforts with new audiences, and participating in projects that help build our region.

On our site, we are focusing on new and cost-effective ways to deliver clean and well-maintained facilities and grounds, well-trained and friendly employees, and memorable experiences for all of our many visitors. Outside our gates, Henry Ford Museum & Greenfield Village works to provide value in several important ways: economic growth in the region through tourism to our campus; educational innovation through pioneering programs like the Henry Ford Academy, our on-site public high school; and "quality-of-life" improvements through participation in a variety of valuable programs and projects—such as the Automobile National Heritage Area and the Rouge River National Wetlands Renovation Project—designed to make metropolitan Detroit an even greater place to live and visit.

Enjoy our history through this volume. Our work, yesterday as well as today and in the future, is very much about providing opportunities for you to rekindle memories, find connections between your past and present, and, through experience and learning, enrich your future. Come see for yourself what Henry Ford and all the rest of us have been doing here since 1929.

Steven K. Hamp
President,
Henry Ford Museum &
Greenfield Village

Steven K. Hamp is shown addressing students at Detroit's Harms Elementary School in 1998.

Henry Ford learned about machinery by working on broken watches and clocks. They were among his favorite old things to purchase—he tinkered with them his entire life. By 1923, when this photo was taken, he had assembled a sizeable collection in the Engineering Laboratory at Ford Motor Company.

Henry Ford's "history is bunk" statement drew immediate press attention. In fact, most Americans continue to share his disdain for the academic "facts" of history and theories of historians.

HISTORY MERE "BUNK"

Ford Simply Regards It All as Tradition

He Calls Professional Soldiers Murderers

Would Not Exclude Pershing or Grant

For League Without Delay or Another War

Mount Clemens, Mich, July 15—Testifying today in his $1,000,000 libel suit against the Chicago Tribune, Henry Ford admitted ignorance of history and said that more than ever he considered it "bunk" growing out of tradition; nor had he any

Henry Ford's ideas about history, education, and his personal legacy inspired him to create a unique collection of things and a different kind of educational institution.

In 1919, Ford uttered what has become one of the most repeated phrases in American history. Henry Ford had sued the *Chicago Tribune* for libel. The paper had called him an "anarchist" and "ignorant idealist" after mistakenly accusing his company of failing to support workers who were called up for National Guard duty fighting Mexican revolutionary Pancho Villa. During the trial, the *Tribune* lawyers examined Ford on a variety of schoolbook topics, history in particular. Finally, Ford expressed his contempt for the emphasis on memorizing facts. Ford, who was not "book smart," testified, "History is more or less bunk. It's tradition. We don't want tradition. We want to live in the present, and the only history that is worth a tinker's damn is the history we make today." The simplistic epigram "History is bunk" stuck in the public mind.

The truth is that Henry Ford never really believed "history is bunk." He fervently believed that history taught from books that omitted ordinary folks and the tools of everyday life was truly useless, as he revealed in 1932:

"When I went to our American history books to learn how our forefathers harrowed the lands, I discovered that the historians knew nothing about harrows. Yet our country had depended more on harrows than on guns or speeches. I thought that a history that excluded harrows and all the rest of daily life is bunk and I think so yet."

Well before he uttered his infamous remark, Ford was assembling

THE FOUNDATION

his own history text by collecting the objects that ordinary folks had designed, produced, and used during the previous two centuries. This early accumulation included plain old wagons, grease lamps, and threshing machines—the kinds of things few other millionaires gave a second thought. Ford believed these objects told the truth not reflected in written histories, professing the belief "in these collections is true history. These relics of days that are gone do not lie." These relics, first collected about 1912, illustrated what Ford saw as "true history."

What kinds of things did Ford begin to collect, and what did they say about his peculiar view of history? Ford consistently sought domestic furnishings used by common folk, objects that reflected industrial progress, transportation vehicles, and some relics associated with great men in American history. Henry Ford gave homage to his heroes Thomas Edison, Abraham Lincoln and educator William Holmes McGuffey by collecting relics of their lives and examples of their work. These seem remarkable historical acquisitions for a man who had said "history is bunk."

Thus, early on, Ford packed away everyday artifacts—some of them manufactured just a few years earlier—such as mousetraps, kitchen chairs or knitted winter bonnets. He sought objects of industrial progress, such as the power thresher and boiler manufactured in 1909 that he purchased from the New York World's

While Henry Ford hated farm work, he loved the family homestead in Dearborn (shown below in 1924). Henry and Clara regularly returned to the homestead to enjoy the rural atmosphere.

There are many stories of Henry Ford's endeavors to re-create the original ambience of his birthplace. Archaeology conducted on site revealed shards of a red transferware plate, and Ford sent pieces to England for reproduction. Those replicas (above) sit in the pantry of the Ford Birthplace today.

Fair in 1912. In the same year, Ford began methodically collecting the precursors to his Model T, including a Conestoga wagon, horse-drawn coaches and his old 999 racing car. Ford's collection illustrated a saying that he adopted as his approach to change: "Mankind passes from the old to the new over a human bridge formed by those who labor in the three principal arts—Agriculture, Manufacture, and Transportation."

Ford first preserved an entire building with the restoration of his beloved birthplace in 1919. A new road forced the clapboard house just 200 yards from its original location; Ford decided not only to move it, but also to restore and refurnish it as authentically as possible to his boyhood recollections. Ford worked for seven years tirelessly seeking original furnishings or exact reproductions of furnishings or equipment. The house was dedicated to the memory of his mother, Mary, and visitors were awestruck by its authenticity and felt as if Mary was a looming presence. Ford brought friends, family and reporters to the house and used it as a pulpit to preach the lessons his mother had taught him—the love of family, the value of hard work, learning "not from school books but from life," and to trust one's intuition.

In fact, other Americans were caught up in the preservation craze of the early twentieth century, striving to preserve sites of bygone days. Attendees to the Centennial Exposition of 1876 in Philadelphia were mesmerized by the exhibited "antiquities" of colonial days, and many came home thinking about collecting "old time" antiques and restoring colonial homes. This is still referred to as the Colonial Revival movement. Historical societies were busy preserving buildings and collecting the papers of local dignitaries. The Society for the Preservation of New England Antiquities (SPNEA) was founded in 1910 to preserve significant, primarily colonial, northeastern buildings and contents on their original sites.

This activity coincided with national concern for the impact of industrial change on urban life, and the influence of vast numbers of southern and eastern European immigrants. These years also marked the Ford Motor Company's most energetic efforts to "Americanize" immigrants by teaching the English language and enforcing middle-class morals and values. It also coincided with Ford's political activities: promoting pacifism during World War I and unsuccessfully running for the U.S. Senate in 1918 with the encouragement of President Woodrow Wilson. Ford collected objects that he felt would convey traditional Anglo-Saxon values to new generations of Americans, regardless of where they were born.

The extensive press coverage of the birthplace restoration provoked a deluge of pleas for assistance with other preservation projects. In 1923, Ford was asked to assist in the restoration of the Wayside Inn of Sudbury, Massachusetts, built around 1689. Henry Wadsworth Longfellow's *Tales of a Wayside Inn* (1863) had made the tavern famous.

After he restored his old homestead, Ford was inundated with requests for assistance for renovation of other buildings. In 1923, he purchased the Wayside Inn, built in the late 1600s in Sudbury, Massachusetts. The Inn featured a restaurant and a venerable collection of New England furniture owned by famous colonial Americans.

Then & Now
Eagle Tavern

The Clinton Inn as it looked in 1925 on its original location (top). Built in Clinton, Michigan, around 1832, this building had served as a stagecoach stop for travelers on the Detroit-to-Chicago road.

The restored Clinton Inn (below) was reassembled in Greenfield Village in the spring of 1929, but left on skids until Henry Ford decided just where its permanent location would be. Once Ford made up his mind, a foundation was built for the building. A large dining room was added to the back of the building to accommodate students and visitors. In the 1980s, the inn was renamed the "Eagle Tavern"—the name by which it was known in the 1850s when Calvin Wood was the proprietor.

Ford immediately agreed, explaining that he felt a debt of gratitude to Longfellow because the author's *Psalm of Life* (included in McGuffey's Readers) was among his very favorite inspirational verses.

Ford purchased the Inn for $65,000 and restored and refurnished it at significant expense. He also bought up much surrounding land, restored two one-room schoolhouses nearby and even diverted the Boston Post Road to keep those Model Ts from ruining the bucolic setting. Interestingly, in 1926 the press crowed that Ford was to "build an old time town" that would preserve the "simplicity and quaint beauty" of the area. No one is quite sure why the expansion never materialized.

Henry Ford consulted his wife, Clara, on the Inn's menu, and she and Mr. Ford stressed good service and quality food. But more important was the educational experience the Inn offered. Ford stated that he was "interested in maintaining at the Inn the daily ways of life for the instruction of school children." Ford thought history came alive when he gazed upon furniture of the great Colonials such as John Hancock or George Washington and heard about their lives. So he filled the Inn with various furnishings of these great men. The staff recorded tour comments and observations and reported that children were entranced—even reverential—when these pieces and their owners' lives were discussed. Israel Sack, the renowned New York City antiques dealer who supplied Ford with many of these pieces in the 1920s, recalled that he would sell "the most beautiful" or "the rarest" furniture to industrialist Henry Francis DuPont

Henry and Clara Ford loved "old time" fiddle music and revived old tunes and contradancing. By the early 1920s Mr. Ford had purchased crude, home-made violins as well as extraordinary Stradivarius violins like the one shown here (above). For a time, before he created the museum, Henry Ford kept his expensive violins in a closet in his home, Fair Lane.

Henry and Clara Ford, fourth and fifth from the right in the front row, liked traditional music and dancing. They often hosted dances with guests in period costume at the Wayside Inn, shown here in 1924. Ford even hired a dance master, Benjamin Lovett, and a band to promote traditional music. Henry and Clara also hosted dances at the Botsford Inn, 16 miles northwest of Detroit, and in the ballroom of the museum's Education Building (also known as Lovett Hall).

for his Delaware furniture museum, Winterthur. Ford was far more interested in furniture "with a history."

Ford knew little about this era or region but jumped into the subject with both feet—learning by doing. Eschewing experts, Ford dispatched Harold M. Cordell, one of his secretaries, to New England to conduct research. Ford himself visited other restored New England buildings such as the Harrison Gray Otis House in Boston. From the Otis House, he lured away an employee named W.W. Taylor to assist him with furnishing the Wayside Inn. This old New Englander could locate anything—he turned up seventeenth-century mills, old stagecoach bugles and ox yokes. But Taylor could also "clinch the deal"—he earned the trust of the rural New Englanders with whom he negotiated purchases.

Reverend William Goodwin of Bruton Parish in Williamsburg, Virginia, wrote to Ford in 1924 urging him to "purchase" the Colonial town of Williamsburg for about $4 million. Ford declined the offer. However, John D. Rockefeller accepted Goodwin's proposal and funded the restoration of Colonial Williamsburg, which opened in 1930. For years these two industrialists corresponded about their private restoration projects, often visiting one another's restored villages. The Fords visited Colonial Williamsburg incognito as "J.H. Jones & Wife." Rockefeller marveled at the service he received at Ford's restored Wayside Inn in Sudbury and later sat for a tintype portrait on a visit to Greenfield Village.

In 1924 the Fords turned to projects closer to home. They restored the little red brick schoolhouse Henry

By 1928, historic objects were arriving in Dearborn in small packages and train-car loads. Ford poses here with a small acquisition, along with two men long influential in his collecting: private secretary Frank Campsall and lawyer Charles Newton.

Looking eastward, this 1927 aerial view (below) shows the Ford-dominated area where the museum and village would be located. Ford had his office in the Engineering Building. He stored his growing collections in an area of the Engineering Building also known as the Experimental Laboratory and in portions of the adjacent tractor assembly plant known as Building 13. The floor plan (right) of the Laboratory and Building 13 shows areas where "Antiques" were stored. Clara Ford would often pull a few of her favorite objects out of cramped storage, then arrange them neatly in a corner of the building (far right).

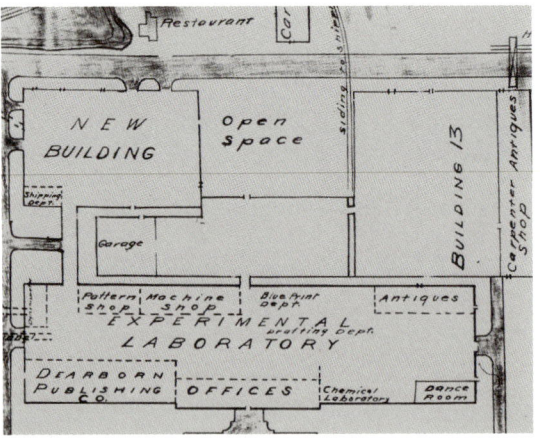

once attended and re-established a one-room school program. The following year they purchased the Botsford Inn, not too far from Dearborn. While Ford did not have any association with this Michigan inn, he loved old-time music and was eager to hold dances there. Ford's staff sought out those old-timers who remembered the Inn in its heyday and used their recollections to help refurnish the place. Henry tracked down the Inn's old fiddler and coaxed him out of retirement in order to fill the Inn with music once again.

As Americans learned Ford was seeking "authentic artifacts," his office was swamped with letters offering "the oldest" or "rarest." Harold Cordell was chiefly responsible for handling this voluminous correspondence; he reportedly looked at virtually every letter lest he miss the opportunity of collecting an extraordinary treasure. His intuition told him what was needed for restorations, what appealed to Ford personally, and what had historical significance.

By 1924, Ford had amassed far more objects than even he had plans for. A huge variety of goods—including carpets, glassware, shoes, cars, clocks and furniture—was shipped by rail and stored rather haphazardly in Ford Motor Company buildings. Cordell recalled that one day Mrs. Ford noticed several beautiful pieces of furniture, clocks and spinning wheels piled up most inelegantly in this storage building. She asked Ford's men to pull out specific pieces and set them up in another end of the building that was more nicely appointed. Here Mrs. Ford arranged room settings that were more artistic than historic, as Mrs. Ford claimed

One of Ford's earliest auto acquisitions was the *999*, a race car that brought him fame and fortune. Ford, standing, drove the car to triumph in a race in 1902. Famous driver Barney Oldfield, seen in the car, set a speed record in it in 1904, thus drawing attention to Ford's manufacturing endeavors.

she knew nothing about antiques. Mrs. Ford seldom gave tours of her mini-museum, but Ford and his staff were eager to show it to interested parties.

After the 1919 libel trial, Ford said to his secretary, Ernest Liebold: "I'm going to start up a museum and give people a true picture of the development of the country. That's the only history that's worth observing, that you can preserve in itself. We're going to build a museum that's going to show industrial history, and it won't be bunk! We'll show the people what actually existed in years gone by and we'll show the actual development of American industry from the early days…up to the present day."

Soon his son, Edsel, was touring museums on the East Coast to see how they were organized and how they presented artifacts and history. In 1923, Henry Ford visited the Mercer Museum in Doylestown, Pennsylvania. Henry Mercer had accumulated a massive collection of exactly the kinds of everyday things that interested Ford. After the visit, Ford said, "This is the only museum I've ever been sufficiently interested in to visit. Some day I expect to have a museum which will rival it." In early 1925, Ford publicly announced his plans to build a museum.

Ford himself made all the decisions about the organization of his collections. Cordell spoke to Henry and Edsel Ford about visiting other museums, and Edsel encouraged Cordell's visits to Chicago's Rosenwald Museum (now the Museum of Science and Industry). Cordell discussed the Deutsches Museum in Munich,

One of Ford's great heroes was William Holmes McGuffey, who created children's school books that included verses and stories of great deeds. In 1913, Henry and Clara Ford began collecting McGuffey's Readers, eventually reprinting them so that children in the Edison Institute School could use them.

which was well organized by subject. Perhaps Cordell even mentioned Skansen, a Swedish open-air museum founded in the late 1800s that emphasized folk life and culture—the kind of everyday life history Ford was so intent on presenting.

In many ways, Ford did not look to other museums as models for his museum. He wanted something more than classification and cataloguing for his new Edison Institute. While Ford listened to Cordell's descriptions of other places and his pleas for planning and classification, Ford knew those things were not important to him. He was more interested in teaching children than in abiding by conventional museum practices.

Ford knew he learned best while using his hands, his intuition and his keen powers of observation. He believed others might learn best in these ways, too, and ensured that his new museum was full of activity and wonder. His museum included the anonymous and the famous, the traditional and the innovative. Like his beloved McGuffey's Readers, Ford's outdoor museum was eclectic—he took the best of the ordinary and extraordinary works of America's people and used them in his "animated textbook" (as Greenfield Village was referred to in the 1930 guidebook) to teach, inspire, and fascinate. Only then would children remember its lessons. ■

The parlor of the Ford Birthplace as it looks today. The entire house was recently re-installed to make it resemble as much as possible its appearance in 1876, when Henry Ford was 13 years old.

As Henry Ford looks on, Thomas Edison signs his name in a block of wet concrete that became the cornerstone for Ford's new museum. The ceremony took place on September 27, 1928. By the following October, the block of concrete was completely surrounded by a 13-acre museum building.

Today, the museum cornerstone is in the main entrance hallway. In addition to Edison's signature and footprints, the concrete slab includes a garden spade once owned by botanist Luther Burbank. The monument represents Henry Ford's vision of the union of agriculture and industry for the betterment of humanity.

In 1926, Henry Ford was already 63 years old. He was famous and rich, and his various restoration projects had made him something of a pioneer in historic preservation.

Having decided to put his collection to educational uses, Ford faced several practical issues. What should he name it? Where would his museum be situated? How would it be designed and laid out? How would the collections be organized and displayed? For whom was the institution intended?

For Ford, the last question was the easiest to answer. He was already supporting schools and colleges around the country. He was convinced that young people learned best by active involvement and direct participation. His "educational museum" was conceived of first and foremost as a school. Neither the general public, tourists nor museum curators were on his mind.

In his plan to create an educational institution dedicated to "learning to do by doing" and the power of artifacts to convey ideas, Ford's major decisions about display were influenced by his philosophy of teaching. There would be an indoor museum where artifacts were arranged by type so that students could see the evolution of progress in any particular technology (steam engines, wagons) or human activity (cooking, tending crops). And there would be an outdoor museum filled with buildings that were artifacts in their own right as well as contexts for the use of other artifacts. Thus, explained Frank Campsall, Ford's assistant, "articles from the Collections of the Museum will be shown in actual use in the Village. Mr. Ford dislikes mere 'dead' exhibitions of things; he wishes to see them in action in their proper setting. Hence the Village."

THE CREATION

The organization was chartered as The Edison Institute, composed collectively of Greenfield Village, The Edison Institute Museum (originally called simply The Industrial Museum) and the Edison Institute School. With Henry, Clara, and son Edsel as the three officers of the nonprofit corporation, the organization's goals were:

"(a) To assemble and exhibit, publish and disseminate historical, scientific, sociological and artistic information and to do any and all things calculated directly or indirectly to advance the cause of education, whether general, technical, sociological or aesthetic.

(b) To demonstrate, for educational purposes, the development of American arts, sciences, customs and institutions by reproducing or re-enacting the conditions and circumstances of such development in any manner calculated to convey a realistic picture thereof."

If only it had not been named Greenfield Village, implying that it was a real, historical place. If it were simply Greenfield Museum, generations of visitors would not have been convinced that Ford, Edison and the Wright Brothers actually lived next to one another in Michigan, to note one of the occasional misconceptions of visitors. True enough, Ford did not collect a jail or a bank or any of a number of other types of buildings, nor did he collect significant buildings from west of the Mississippi. His goal was not to show absolutely everything found in an American town nor illustrate the entire American experience. He focused on illustrating the

Robert O. Derrick (left), the architect chosen by Henry Ford to design his new museum, was one of the few people working on the museum project who was actually professionally trained for the job.

Derrick's original plan for Henry Ford's museum (below) included a second-floor exhibition area and a domed roundhouse at one end, where the architect envisioned historic locomotives would be displayed.

Actual construction of the museum began in April 1929, with major attention being paid to completing the center section in time for the planned October opening.

July 1929 (top) — Scaffolding surrounds most of the Independence Hall replica. Derrick matched that building's original exterior dimensions precisely, but used modern materials like steel girders and reinforced concrete.

August 1929 (center) — Curved steel window frames lie stacked in front of the Independence Hall replica while workmen labor on the wings.

October 2, 1929 (below) — Nineteen days before dedication: The newly painted clocktower awaits the arrival of a reproduction of the Liberty Bell (cast at the same foundry as the original) and a Seth Thomas clock.

July 1929 — While the front of the museum was coming together in good order, the back section proceeded at a much slower pace. Square footings with steel reinforcing rods can be seen, which eventually provided the base for the columns that support the roof above the 8-acre exhibition hall.

August 1929 (opposite, left) — Inside the Independence Hall replica, plasterers work on the massive ceilings while other contractors lay large concrete tiles that will serve as the sub-floor.

November 1929 (opposite, right) — One month after the dedication dinner, the back of the museum was still under construction. It would be many more months before the 8-acre exhibit hall was completely enclosed, and several years before the structure was finished.

values of independence, resourcefulness, and ingenuity that he felt best contributed to America's heritage and welfare.

Besides, students in the Edison Institute School read the textbooks of the day to learn of politics and wars; but when it came to understanding life as lived by earlier Americans, they employed the resources of the museum and village. Ford was assembling a collection of primary sources accessible to inquiring students, not a textbook or an encyclopedia.

Henry Ford had become the primary collector of Americana in the world. The objects sent to Ford by well-wishers, as well as those obtained from Ford's own searches, filled to overflowing the old tractor factory he was using as a warehouse. "It is not as yet an ordered collection," Ford said in 1926. "We want to have something of everything….One of these days the collection will have its own museum at Dearborn, and there we shall reproduce the life of the country in its every age."

Henry Ford and Edward J. Cutler, the man Ford selected to be the architect for Greenfield Village, scouted out several potential sites for Ford's museum complex. Cutler later recalled how he and Henry Ford selected the present site in 1926:

"It was just a field. Then there was an old road that went through there with some old trees and a couple of houses on it, one on top of the old knoll there, where Jimmy Humberstone lived. We walked up the old road. Then he [Ford] took me over to…the end of Cherry Hill Road where the golf links are now, and we walked all over that property ….There wasn't the chance for development over there, and it was separate from all the activities of the company. This present location was much more favorable. I know he liked that better."

The present site was right in the middle of where Ford lived and worked. The museum and village would be close to the engineering building where Ford kept his office, within sight of Ford Field with its Tri-Motor airplane factory, and a short drive away from Fair Lane, Henry Ford's estate. The building site was also close to the Rouge River and had a railroad siding running through it.

Ford saw no contradiction in locating his historical collection

Menlo Park, New Jersey, 1880

Menlo Park

The centerpiece of Greenfield Village was Ford's faithful reconstruction of the Menlo Park Laboratory Complex, the site in New Jersey where Thomas Edison created some of his most famous inventions, such as the phonograph and the light bulb. Little remained of the original site in 1928, so Ford began reconstructing the six buildings in Michigan using old drawings, photographs, and memories of some "old-timers," including Edison himself.

Thomas Edison was reportedly somewhat uneasy about having the buildings of his Menlo Park Laboratory Complex standing so far from their native New Jersey. Henry Ford responded by hauling in seven rail cars of New Jersey clay to spread over the compound.

The Greenfield Village installation includes the more than 2,000 bottles of chemicals Edison kept at the lab (opposite). Shown in the foreground is an early phonograph.

Menlo Park, Greenfield Village, 1929

amidst his modern industrial empire with its airport, engineering labs and massive Rouge River auto manufacturing plant. The contradiction would have been for Ford to find a peaceful, isolated place to hide his historical preservations. For Ford, the past was part of the present and the present was simply the history people were making now and would be studied through their artifacts in 20 or 50 years. As he said, he collected "up to the present…the history of our people as written into the things their hands made and used."

While Cutler was busy locating historic structures, Ford enlisted the services of architect Robert O. Derrick to design the building that would house his museum of Americana. Derrick suggested that Ford include a replica of Philadelphia's Independence Hall, an idea that Ford immediately endorsed. Derrick studied the designs of several museums and drew up some detailed plans. He originally envisioned a massive building with a number of courtyards breaking up the single expanse.

Derrick recalled that:

"He [Mr. Ford] said we would have to have a model made, so we had a model made and it showed the balconies, naturally, and the basement, and he said, "What is this up here?" I said, "That is a balcony for exhibit." He said, "I wouldn't have that; there would be people up there, I could come in and they wouldn't be working. I wouldn't have it." He said, "I have to see everybody." Then he said, "What's this?" I said, "That is the basement down there, which is necessary to maintain these exhibits and to keep things which you want to rotate, etc." He said, "I wouldn't have that: I couldn't see those men down there when I came in. You have to do the whole thing over again and put it on one floor with no balconies and no basements." I said, "Okay," and I went back and we started all over again. What you see now is what we did the second time."

Henry Ford's architect for Greenfield Village, Edward Cutler, worked in the glass department of Ford Motor Company before being asked to design a wooden windmill for Ford's boyhood home.

Derrick's revised plan consisted of a massive front façade featuring replicas of Independence Hall, Congress Hall and Philadelphia's City Hall. Derrick created a 350,000-square-foot exhibition hall, an area the size of seven football fields. This gigantic room on one level was reminiscent of contemporary factory design: a grid of supporting columns approximately 40 feet apart, each column carrying hot water radiators for heating, and skylight windows to catch the sunlight. To some, the symbolism was unmistakable. The architecture of The Edison Institute Museum embodied Ford's reliance on progress through political independence and modern technology.

The cornerstone for Henry Ford's new museum building was officially christened by Thomas Edison on September 27, 1928. Edison had come to Michigan to start up the steam engine in his old laboratory that Henry Ford had moved from Fort Myers, Florida, and reconstructed as the first building in Greenfield Village. Ford's complex seemed to be off to a modest start. And yet, just 13 months later, Edison would return to find that the village had mushroomed in size and now contained 28 buildings, including a reconstruction of his own Menlo Park laboratory compound. Ford's huge museum building was going up, immediately adjacent to the village. Why the rush?

A major anniversary was coming. October 21, 1879, was acknowledged as the date when Edison successfully illuminated his first electric lamp, and a number of groups looked forward to celebrating the fiftieth anniversary in 1929 in an event to be called "Light's Golden Jubilee." A group of Edison's friends and former associates, called the "Edison Pioneers," were planning an elaborate public festival, but realized that they lacked sufficient funding. They approached the General Electric Company to take over the entire program. General Electric had long ago absorbed

Between 1928 and 1929, while hundreds of workmen labored to build the 13-acre museum building, hundreds more were at work creating Greenfield Village.
By the time of the dedication ceremonies on October 21, 1929, the village included 28 historic structures. By 1933, when the site was opened to the public, Ford's village had over 50 buildings in place.

Then & Now
Sir John Bennett Jewelry Shop

Gog and Magog, two mythical British giants, tolled the Westminster Chimes each quarter hour from 1846 to 1929 above Sir John Bennett's Jewelry Shop in London, England. When Henry Ford learned that the building was going to be demolished, he purchased the structure and shipped the clock mechanism with assorted building parts to Michigan.

Once in Greenfield Village, Edward Cutler trimmed the Sir John Bennett Shop from its original five floors down to a smaller two-story structure that has become a village landmark.

Edison's lamp business and saw the upcoming anniversary as a great public relations and sales opportunity.

However, no one at General Electric had consulted with Edison in advance about the celebration; he was merely informed as to when and where he was to appear. Neither Ford nor Edison saw any reason why Edison should provide so much publicity and free advertising to General Electric. In early 1929, as Edison biographer Matthew Josephson describes it:

"…he [Henry Ford] suddenly appeared in the library of his venerable friend at West Orange, and waited for him, walking up and down restlessly, and muttering to himself: 'I'll show 'em. I'll kidnap the whole party.' Edison agreed to be 'kidnapped.'"

With Edison on board, Ford convinced General Electric to participate as a junior partner in the Jubilee, publicizing the celebrations around the world. Once Ford decided that the dedication of his museum complex would coincide with the celebration of Light's Golden Jubilee, the pace of construction accelerated dramatically. Fortunately, as one of the world's richest men, Henry Ford had access to all the resources he needed to complete the job. By October 21, 1929, the front sections of the museum building were sufficiently completed so that the dedication dinner could take place. The reconstructed Menlo Park laboratory complex provided the perfect backdrop for the light bulb re-enactment.

Ford had succeeded in kidnapping Light's Golden Jubilee away from General Electric and had made it his own. Ford's new museum complex, formally named The Edison Institute in honor of his friend and mentor, was properly christened, with dignitaries from all over the country in attendance—including the President of the United States and the president of General Electric. ■

The honor of your presence is requested by Mr. Henry Ford and Mr. Edsel Ford at a Celebration in honor of Mr. Thomas Alva Edison on the occasion of The Fiftieth Anniversary of his Invention of the Electric Light and the dedication of The Edison Institute of Technology by The President of the United States on Monday, October twenty-first Nineteen hundred and twenty-nine Dearborn, Michigan

Prior to the dedication of The Edison Institute and celebration of Light's Golden Jubilee, some 500 guests gathered at Smiths Creek Depot in Greenfield Village to welcome the Edisons, the Hoovers and the Fords. The hosts and guests of honor arrived on "The President," a wood-burning, 1850s locomotive with baggage and passenger cars commemorating Edison's work on Michigan's Grand Trunk line 67 years earlier.

The RSVPs for Light's Golden Jubilee began pouring in to Ford Motor Company by early October 1929. While business tycoon John D. Rockefeller was unable to attend, his son, John Jr., accepted in his place. J.P. Morgan and Orville Wright would attend. Madame Marie Curie graciously accepted and so did Will Rogers in his own humorous way, adding on his acceptance telegram that he "hopes [President] Hoover makes a good speech."

The morning of the dedication of The Edison Institute and the celebration of Light's Golden Jubilee brought forth the rains. Twenty-eight buildings were assembled in Greenfield Village from around Michigan and the United States for the dedication, with Edison's Menlo Park, New Jersey, laboratory as the centerpiece. The muddy grounds made maneuvering and sightseeing around the massive village a trying task. However, the rains didn't dampen the enthusiasm of those present.

At 10 a.m., President Herbert Hoover, Henry Ford and the guest of honor, Thomas Edison, arrived at Smiths Creek Depot at Greenfield Village on a commemorative steam-powered locomotive, much like the one on which Edison had sold papers as a youth. More than 500 invited guests, under a sea of umbrellas and rain slickers, roared their approval and congratulations as Hoover came off the train, arm-in-arm with Edison with Henry Ford close behind. All three stood on the same platform where Edison had landed some 70 years earlier, after being thrown off the train for starting a fire. Edward L. Bernays, one of the early pioneers of modern public relations and the man who helped General Electric promote the historic year, quickly

THE DEDICATION

organized the guests to greet Hoover, Edison and Ford.

After the guests had been properly greeted and the throngs of media got their quotes and pictures, Ford gave Hoover a personal tour of the Rouge Plant, five miles away. Eighty-two-year-old Edison retired to nearby Fair Lane to rest before the dedication. The hundreds of guests on hand gathered at the Clinton Inn in the village for lunch and afternoon horse-and-carriage tours through the village.

That evening, The Edison Institute was literally aglow. English-made crystal chandeliers lit the room. To commemorate Light's Golden Jubilee, Ford had these fixtures fitted with tallow candles, hand dipped that day in the village. They were to be extinguished and massive floodlights would light up The Edison Institute, but only after Edison successfully re-created the lighting of his incandescent lamp. In fact, the unseasoned candles burned so quickly that the back-up electric lighting system had to be used earlier than planned.

NBC Radio broadcaster Graham McNamee was heard coast to coast, narrating the re-creation of the lighting of the first electric light bulb, and he set the mood for the evening to a listening world. "Imagine the checkered effect of black and white evening dress, the brilliant splashes of color provided by the uniforms of military attaches and the great stylists of Paris and Fifth Avenue…I have attended many celebrations, but

The guests, dignitaries and media braved the cold October rains to greet Ford, Hoover and Edison at the Smiths Creek Depot in Greenfield Village. Edison recalled once being thrown off the train at Smiths Creek after accidentally setting fire to the baggage compartment.

To combat the rain and mud, Ford supplied covered horse-drawn carriages to transport guests on tours of Greenfield Village.

Preceding Light's Golden Jubilee, employees dressed in period costumes, like this high-wheeled bicycle rider outside the Sarah Jordan Boarding House, entertained guests in the village.

I cannot recall even attempting to describe one staged in a more perfect setting."

Edison, Ford and Hoover went to the reconstructed Menlo Lab in Greenfield Village. There, Edison met Francis Jehl, his former assistant. They both went to work, much like they had half a century earlier, preparing to forever change the world.

Worldwide publicity of Ford's event caused people in the United States and other parts of the world to huddle around their radios, plunged into near darkness, using only candles or gas lamps for light, waiting for Edison's successful re-creation as a cue to turn on their lights as part of the celebration. McNamee narrated to a hushed world, "Mr. Edison has two wires in his hand; now he is reaching up to the old lamp; now he is making the connection… It lights! Light's Golden Jubilee has come to a triumphant climax."

At that moment, a replica of the Liberty Bell, cast from the same mold, sounded in the belfry at the entrance of The Edison Institute to signal Edison's triumph. Overhead a plane flew by with the word "Edison" and the figures "79" and "29" illuminated under the wings. Car horns sounded, lights flashed on and off, and the world bathed itself in electric light in tribute to Edison.

After the re-dedication, the group returned to the Great Hall of the museum and heard accolades from Madame Curie and President Hoover, and even a radio address by Albert Einstein from Germany, who praised Edison for his contributions to the world.

Then & Now
Logan County Courthouse

Not everyone approved of having local landmarks relocated to Henry Ford's new museum. Ford quietly purchased the old courthouse in Postville, Illinois, where Abraham Lincoln had once practiced law. When local residents learned of the plan, they went to court seeking an injunction to prevent the building from leaving town. Cutler later recalled, "By the time they had their legal end of it taken care of, we had the walls and whole thing flattened to the ground and were carting it off."

Once the courthouse arrived in Greenfield Village, it went up nearly as fast as it was taken down. Dismantling had started in Illinois on September 8, 1929. By September 26, workers were building a foundation for the building in Greenfield Village, facing the village green. Less than a month later, the Lincoln Courthouse (now Logan County Courthouse) was finished.

The event was considered a great success. The world listened in on radio and took part in the festivities as Edison re-invented the light bulb. Telegrams poured in to The Edison Institute from world leaders and other dignitaries offering congratulations. The U.S. Post Office even issued a special two-cent stamp featuring Edison's light bulb.

Edison marveled and was very pleased with the reconstruction of the Menlo Park Laboratory, from the red New Jersey clay that Ford brought in for the foundation to the re-creation of the pipe organ in the lab that Edison's workers played for entertainment.

Those in attendance called the celebration "the event of the century in every respect," and "exceptionally good." The village was described as "unique" and an "original and altogether attractive design." One attendee even thanked Ford in advance "in founding The Edison Institute of Technology for the use and enjoyment of our posterity." ■

Thomas Edison (sixth from the left and wearing a neckerchief) sits with members of his team on the second floor of the Menlo Park main laboratory building in 1880.

The second floor of the restored Menlo Park main laboratory building as it looks today (below). Thomas Edison proclaimed Ford's restoration to be "99.9% accurate" when he saw it in 1929. Why not a perfect 100%? "Because we never kept it this clean," Edison explained.

October 21, 1929

Henry Ford would not allow any flash photography during the Light's Golden Jubilee dinner. To compensate, in 1938 Ford commissioned his staff artist, Irving Bacon, to document the event with a 17- by 7-foot painting of the banquet. Ford even had Bacon add his son Edsel's family, who were ill during the Jubilee and did not attend. The painting was completed in 1945 and currently hangs in the museum.

John D. Rockefeller Jr.
Marie Curie
Herbert Hoover
Henry Ford
Clara Ford
Thomas Edison
Will Rogers
Harvey S. Firestone
William E. Scripps
Orville Wright
Ransom Olds
Charles M. Schwab
Hiram H. Walker

The day after the historic event and dedication, *The Detroit Free Press* gave front-page coverage. The paper remarked that "millions…listened [to the dedication] over the radio in the evening, while Tom Edison by the dim rays of an oil lamp and with the president standing silent and serious in the deep shadows behind him, recreated that first incandescent bulb which was born of his brain 50 years ago."

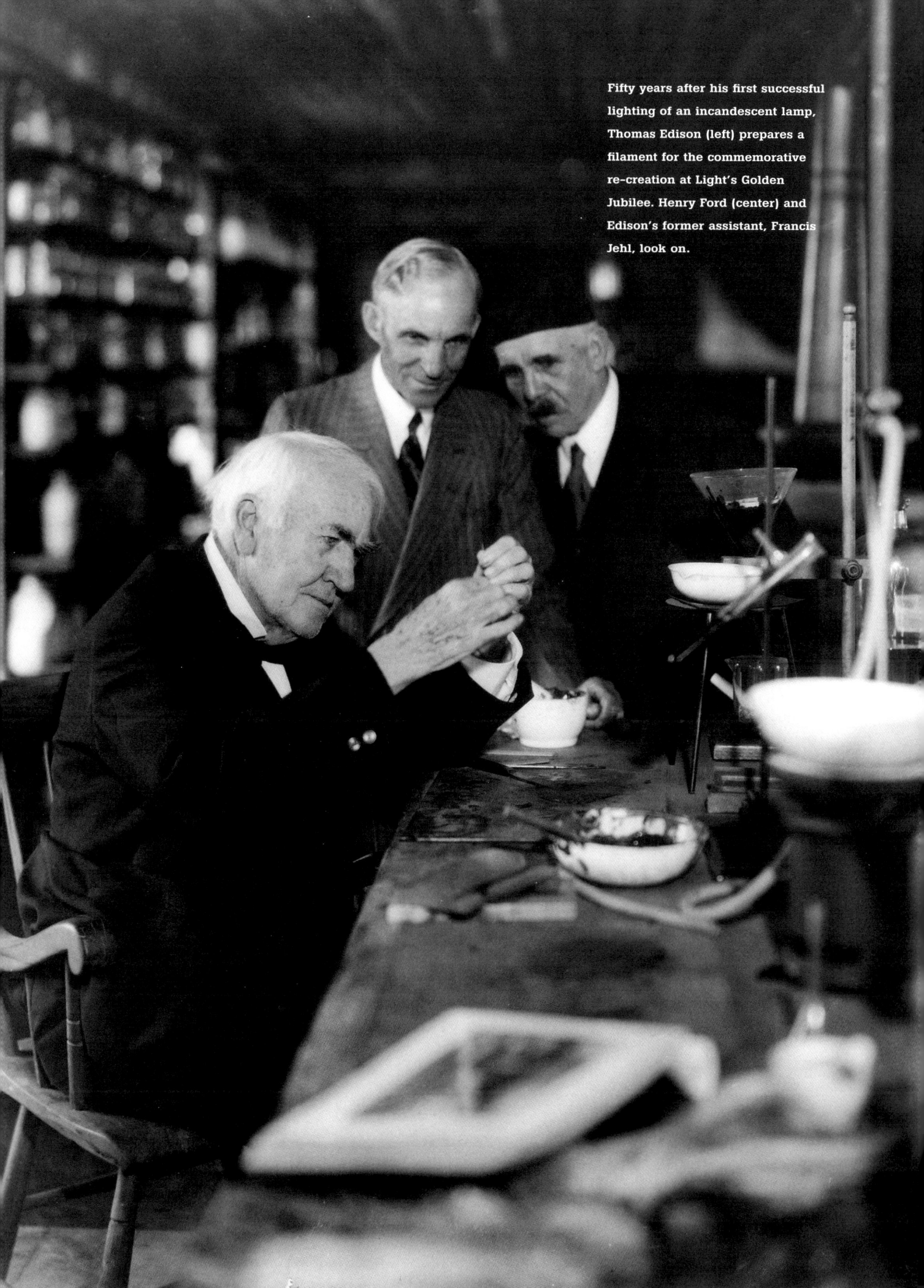

Fifty years after his first successful lighting of an incandescent lamp, Thomas Edison (left) prepares a filament for the commemorative re-creation at Light's Golden Jubilee. Henry Ford (center) and Edison's former assistant, Francis Jehl, look on.

Famous Friends

In 1973, *The Phil Donahue Show* was broadcast live for a week from Town Hall in Greenfield Village (top). Here, Phil Donahue interviews his guests, Motown recording artists Gladys Knight and the Pips.

Actor Jimmy Stewart is shown in 1979 with a bust of aviator Charles A. Lindbergh and the replica *Spirit of St. Louis* (right). Stewart portrayed the famous pilot in the 1957 motion picture based on Lindbergh's autobiography. Stewart, himself an accomplished pilot, was given the replica at the film's completion and later donated it to the museum.

Actor James Earl Jones closely examined the vacuum pump inside Thomas Edison's Menlo Park Laboratory while on a visit in 1993 (below).

Walt Disney, shown here with a student from the Edison Institute School, gathered ideas for his theme park, Disneyland, while visiting Greenfield Village in 1940.

In September 1929, Henry Ford greets students on the first day of school in front of the restored Scotch Settlement School—one of the first buildings brought into Greenfield Village and the same one-room schoolhouse that Ford himself attended as a boy.

The international publicity arising from the October 1929 event, as well as the numerous articles published in national magazines and newspapers, intensified public interest in Henry Ford's historical venture. Visitors—casual or formal, local or foreign—were sometimes offered tours of the village; sometimes they were turned away. No one was admitted into the museum building without a special pass. Students at the Edison Institute School had full run of the place.

However, by 1932, it was evident that the time had come for formal arrangements to accommodate a growing public. Construction began on a colonial-style waiting room, known as the Gatehouse, which would serve as the formal entrance to the village. Staff transported visitors in horse-drawn carriages from the gatehouse to the Clinton Inn, where they began their tour of Greenfield Village. The first formal public visitors were admitted on June 22, 1933. The yet-unfinished museum welcomed its first official visitor about a week later, offering the public the unusual opportunity to see this unique project still in the making. Visitors paid for admission: 25 cents for adults and 10 cents for children. School groups were admitted free of charge. The very next year, 1934, a quarter of a million paying visitors came.

Although Henry Ford had formally opened his museum and village, his quest to locate and preserve historic buildings and artifacts continued. He expanded the roster of American heroes represented in Greenfield Village. The Pennsylvania log home that had been the birthplace of William Holmes McGuffey, the nineteenth-century educator whose textbooks had so influenced Ford as a youth, arrived in 1934.

It is telling that Ford was so taken with collecting McGuffey relics and

THE EXPANSION

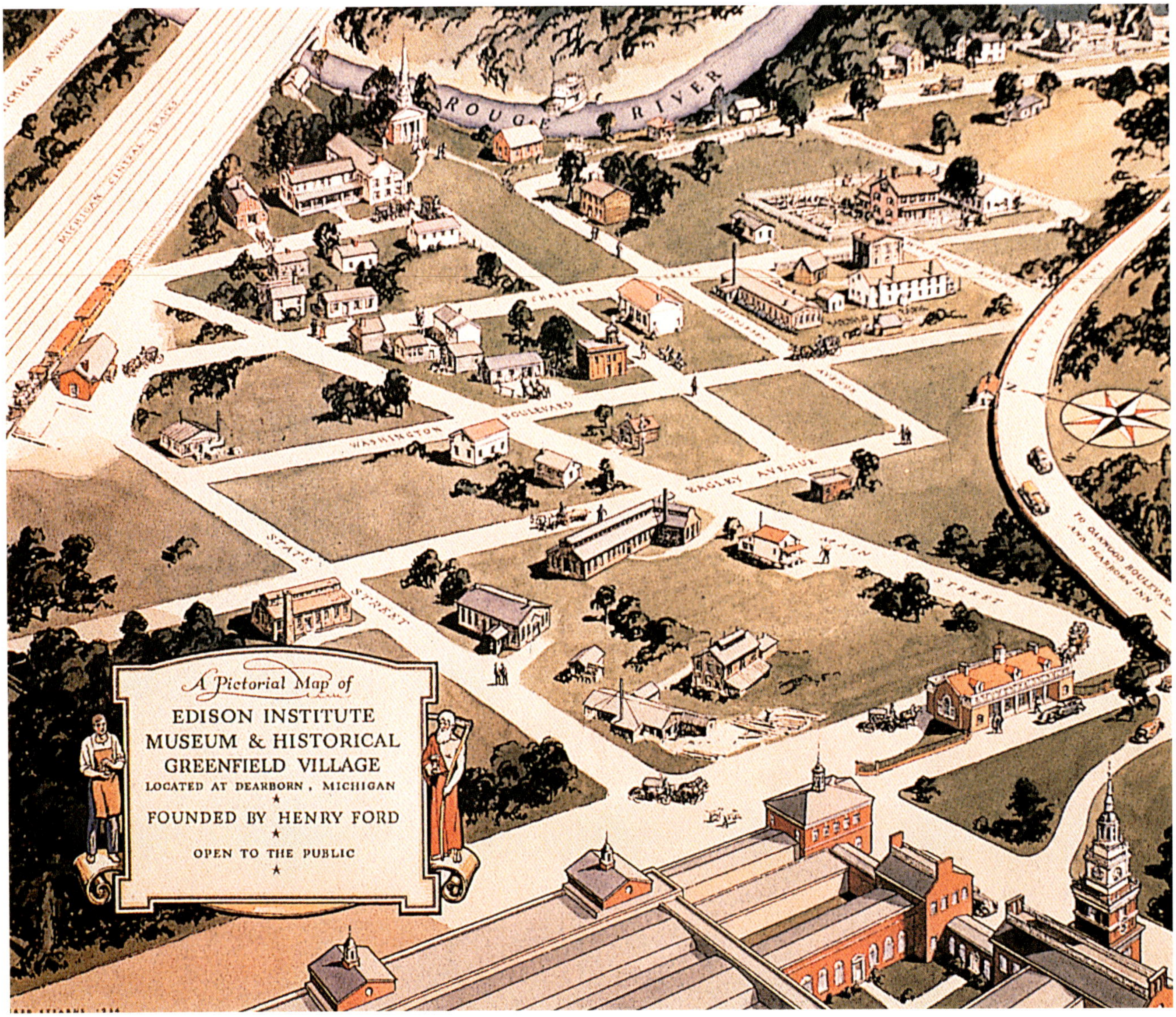

This map shows the village as it looked when officially opened to the public on June 22, 1933 (above). Visitors were transported in horse-drawn vehicles into the village where they began a guided tour of Henry Ford's living history book.

As the number of people requesting tours through Greenfield Village grew to nearly 1,000 a day by 1932, Henry Ford agreed to open the doors officially. Construction began on this colonial-style entrance building in the summer of 1932 (left).

readers. These readers, which Mr. and Mrs. Ford had remembered from childhood, were "eclectic," meaning they included examples of popular and classic poetry and prose in order to teach children traditional values and moral truths as well as reading and writing.

McGuffey had a great sense for what would appeal to children and included dramatic and colorful poetry by authors such as Longfellow, as well as popular children's ditties such as "Mary Had a Little Lamb." Children found them to be great fun as well as instructive and tended to remember these verses from their school days. One afternoon in 1914 Henry and Clara Ford had a grand time recalling beloved passages and decided to find copies of old McGuffey's Readers. The Fords bought every copy they could find, and a few years later even had them reprinted for a new generation of readers.

Ford admired the accomplishments of self-made men, like pioneer aviators Wilbur and Orville Wright. In 1937, he acquired the Wright family home and bicycle shop from Dayton, Ohio. Ford met African-American scientist George Washington Carver in 1936. In 1942, Ford built a small log cabin in tribute to Carver's contributions to agricultural research and education.

As word of Ford's passion for collecting historic homes and workshops spread, buildings began to "seek" Henry. When Yale University planned to demolish Noah Webster's home in New Haven, one of Ford's dealers brought the house's plight to Henry's attention through his son, Edsel. In 1936, the national organization of Ford Dealers presented a seventeenth-century Cape Cod windmill as a gift to Ford for his village. The Susquehanna house from Maryland was

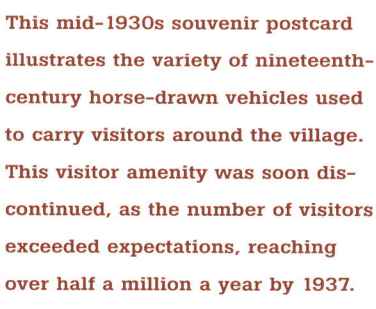

This mid-1930s souvenir postcard illustrates the variety of nineteenth-century horse-drawn vehicles used to carry visitors around the village. This visitor amenity was soon discontinued, as the number of visitors exceeded expectations, reaching over half a million a year by 1937.

offered to Ford when the land on which it was located became a naval airbase during World War II. Thus, the outdoor museum developed the same eclectic nature that distinguished the rest of the collection.

During this time, Henry Ford also began to acquire and preserve buildings relating to his own history. In 1933, he reconstructed the Bagley Avenue shed in which he built his first automobile. Ford also created replicas of one of the schools he had attended, the Edison Illuminating Company where he had been employed early in his career, and the first Ford Motor Company factory. In 1944, Henry Ford moved to Greenfield Village the family homestead where he had been born, although it was not opened to the public until 1953.

Artifacts continued to stream into the collection. Often Ford acquired collections amassed by others, such as the over 200 horse-drawn and motorized vehicles purchased from a California collector. He collected the uncommon: George Washington's camp chest and bed, the chair Abraham Lincoln was sitting in when he was assassinated, and spectacular examples of colonial silver, pewter, and furniture.

Unusual for museums of the era, the collections included items of quite recent vintage like the 1939 Sikorsky helicopter, the first successful helicopter. Many artifacts—automobiles, agricultural equipment, furniture, radios, and domestic appliances—were arranged chronologically in the museum to illustrate Ford's version of the march of progress, right up to the present.

Collecting did not stop when the village and museum doors opened. Ford continued to acquire historic structures and artifacts. The transfer of the Stephen Foster Memorial from Pittsburgh in 1934 prompted *Pittsburgh Post-Gazette* cartoonist Cyrus Hungerford to depict Henry Ford as the "Pied Piper" of historic buildings.

The Owl Night Lunch Wagon provided food and beverages for hungry visitors, serving as the only village refreshment stand through much of the 1930s. During the 1890s, Henry Ford had frequently patronized the Owl whenever he worked the night shift as a young engineer at the Edison Illuminating Company in Detroit.

Then & Now
Cotswold Cottage

Henry Ford wanted an English stone house for Greenfield Village to help show how Americans' ancestors lived. In 1930, Ford's agent found available buildings built around 1620 in the Cotswold region of England. The building was modified by adding a second floor dormer and bay windows to fit Ford's image of an early English cottage.

The cottage and its adjacent barn were disassembled stone by stone, along with a 300-year-old blacksmith shop. In all, over 475,000 pounds of limestone, occupying 67 English train cars (right), were shipped to Greenfield Village. A team of English builders came along with the tons of limestone to assist with the re-assembly of the buildings.

Known today as Cotswold Cottage, the seventeenth-century house and barn, surrounded by a typical early twentieth-century garden, are some of the most popular buildings in Greenfield Village.

Various village buildings served as classrooms for the elementary grades. A photographer caught these young students at their lessons in the mid-1700s Giddings House in September 1944. High school students attended classes in the museum.

Lovett Hall, designed by museum architect Robert O. Derrick and completed in 1937, provided facilities for Henry Ford's growing school system. The building housed a gymnasium, squash courts, a pool and a ballroom, as well as a library, laboratories and classrooms for the Edison Institute of Technology, a college-level work-study program.

Yet all the while, it was his educational purpose—and not merely collecting—that was foremost in Ford's mind. To Henry Ford, the collection was a means to an end, and the school was the priority. He gave the students free run of the village and the museum. The guiding staff was instructed not to let the visiting public interfere with the ability of the students to enjoy their surroundings.

New students were added each year to the 32 who had begun classes in 1929, so that over 270 were attending in grades kindergarten through college in the late 1930s. Students chosen for the tuition-free school came from a waiting list representing a variety of backgrounds and were not necessarily related to Ford Motor Company employees, although nearly all came from Dearborn. A committee decided who would attend, but Ford often added children after personal requests.

Pupils in the Edison Institute School enjoyed the run of the village and museum. They were encouraged to actively learn from the historic buildings and objects surrounding them, and they enlivened the village with their play.

Numerous village buildings and rooms in the museum became classrooms for this growing student body that graduated its first high school class in 1937.

Ford believed that "by looking at the things that people used and that show the way they lived, a better and truer impression can be gained than could be had in a month of reading." In his Edison Institute School, students would learn not only from books, but also from objects. "A piece of machinery or anything that was made is like a book—if you can read it. It is part of the record of man's spirit," said Ford. It was a means of learning that replicated Ford's own experience and the way, in fact, that he himself learned best. Ford was also committed to "functional education" that gave young people experiences with ways of making a living. The key to his philosophy was, as he put it, "Learn to do by doing—that's my favorite principle in education." It was also a highly publicized educational philosophy promoted by John Dewey, a prominent educator at the University of Chicago.

Students were taught using both traditional and progressive methods. Standard academic subjects like reading, arithmetic, geography and science were at the core of the curriculum. Pupils used the artifacts and historic buildings for practical learning. Girls practiced homemaking skills while boys got experience in a machine shop. Reflecting Ford's interest in balancing urban and rural experiences, each pupil worked a vegetable plot in the village.

The younger children made the village their playground—jumping rope, roller skating and sledding among the historic buildings. Students also enjoyed an amazing array of other activities, offering an outlet for whatever talents and interests they possessed. They researched and wrote articles for the school newspaper, acted in plays, learned to ride horses, played on the basketball team, wove textiles, made pottery, and operated a radio station. Ford shared the educational ideas of Elbert Hubbard, a New York advocate of the arts and crafts movement, in believing that, as Ford put it, "Education is not just preparation for life, but a part of life itself…a continuous part."

While the village and museum gave Ford an outlet for his ideas on education, they also provided a refuge from his business concerns. Edward

Ford is shown here with Orville Wright on the porch of the Wright home in 1936. Ford had hoped to acquire the Wright brothers' first plane as well, but it ultimately found a home at the Smithsonian Institution.

Then & Now
Wright Home and Cycle Shop

The Wright family lived in this Dayton, Ohio, home on and off for more than 40 years. Orville was born in the home in 1871, and Wilbur died there in 1912. The spacious porch, said to have been designed and built by the brothers in 1892, is visible in this turn-of-the-century photograph.

Wilbur and Orville Wright built their first airplane in the bicycle sales and repair shop they operated in this Dayton, Ohio, building from 1897 to 1907. In this photograph, taken about 1908, the words "Wright Cycle Co." are barely visible above the shop window.

Information supplied by Orville Wright, as well as by the family housekeeper and a former Wright Cycle employee, helped in the restoration of the Wright home and Cycle Shop. Members of the Wright family, including Orville, donated many of the artifacts displayed in the home and Cycle Shop.

The museum was in various stages of completion when the first visitor walked through its doors in June 1933. Laying the herringbone pattern teak floor was a painstaking process that was not completed until 1938.

Cutler, who oversaw the movement and reconstruction of village buildings, noted:

"[My] office and his work in the village were safety valves for the pressure and strain of the Ford Motor Company…[Ford] spent so much time around the village… It was a relief for him to get down there. For years he wouldn't let me have a telephone. When I would ask him about it, and I had a lot of running around to do, he would say, 'Oh, forget that stuff. I came down here to get away from that gang.' He didn't want any way for them to get a hold of him. I eventually got a telephone, but I had to ask for it a good many times."

The atmosphere of the village in particular appeared to strike an emotional chord with Ford. He delighted in the casual atmosphere of small homes, craft shops and stores. The village was not just a refuge in the past; it was the way Ford thought people should live and work in the present and future. Cars, he believed, would allow people to move out of congested cities. Rural factories would process farm products and allow workers to grow their own crops as well as earn wages. Small communities would encourage individual independence and self-sufficiency. The village was as much his prescription for the present as it was his description of the past.

By the mid-1930s, the number of visitors each year reached half a million. Ford's museum and village communicated the homespun values of a pre-industrial American past, as well as the belief in technological progress. For visitors of Ford's generation, the village and museum brought a wealth of personal memories, as they not only viewed objects from a distant past, but also those that were a part of their everyday lives.

Visitors saw a work in progress in the unfinished museum (below). The decorative arts displays in the front corridors progressed faster than the mechanical arts exhibits in the back of the museum, but arrangements of artifacts were often still temporary. This photograph shows the glass collection as it appeared in February 1935.

Built during the 1930s, the Street of Shops (below, right) was Edsel Ford's idea to link the mechanical arts displays on the back floor with the decorative arts in the front corridors. The small-scale buildings, which featured craft tools, products, advertising materials and folk art, were removed in the 1980s.

Although Ford's museum and village were popular with everyday visitors, these projects had little impact on the development of other historic houses and museums. Uninfluenced by historic preservation or museum philosophies, Ford was not interested in professional standards or scholarship. Preservationists questioned the desirability of removing historic buildings from their original surroundings. Some of the buildings moved to or built in Greenfield Village did not approach the authenticity and accuracy of the Menlo Park project.

Ford's project appeared increasingly eclectic. It was made up of many buildings that reflected Ford's own lifetime or his particular interests. Ford acquired what he liked, and he portrayed a history conveying the values he thought most useful.

And, while technological history was largely ignored at other museums, Henry Ford emphasized machinery and other technological inventions. This emphasis on social and technological history was ahead of its time. Not until the 1960s would historians and museums begin to investigate and interpret the everyday life of common people.

"This is the only reason Greenfield Village exists—to give us a sense of unity with our people through the generations, and to convey that inspiration of American genius to

A cherished tradition began in June 1935 when the first wedding was held in the Martha Mary Chapel. Since that time, thousands of couples have taken their vows in the chapel in Greenfield Village.

Edsel Ford, George Washington Carver and Henry Ford stand outside the George Washington Carver Memorial, built in Greenfield Village in 1942 to honor the world-renowned agricultural scientist. Carver was born into slavery in a log house like this one.

In 1943, Igor Sikorsky landed his 1939 Vought-Sikorsky helicopter on the front lawn of the museum, before thousands of schoolchildren.

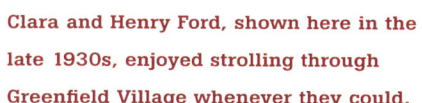

The onset of World War II created a shortage of the young men who had served as village and museum guides during the 1930s. Beginning with the 1942 summer season, high school and college students attending the Edison Institute School were recruited for the guiding staff. Two years later, students from other local area high schools joined them.

Clara and Henry Ford, shown here in the late 1930s, enjoyed strolling through Greenfield Village whenever they could.

our youth. As a nation we have not depended so much on rare or occasional genius as on the general resourcefulness of our people. That is our true genius, and I am hoping that Greenfield Village will serve that."

Saying this, Henry Ford rejected the approach of glorifying famous patriots and the highborn. He chose instead to represent the common man—farmers, blacksmiths, machinists—or self-made men like Edison and the Wright Brothers.

As Henry Ford's health began to fail in the 1940s, the pace of activity in the village and museum slowed. In 1943, his son, Edsel, who was an able and enthusiastic supporter of arts, culture and education, died at the early age of 49. Few historic buildings were added, and even Ford's beloved school system was reduced in size during World War II. When Henry Ford died at age 83 in April 1947, the future of the village and museum was very much uncertain. ■

In January 1944, Ford's birthplace was simply cut in two and hauled to the village by truck (below). The home's proximity to Greenfield Village meant that it did not have to be disassembled. The house joined other village buildings relating to Henry Ford's own history.

When Henry Ford died, his body lay in state in the lobby of Lovett Hall. More than 100,000 people waited in line (shown above in front of the Village Entrance building) to pay their respects.

Famous Friends

Edsel Ford, Henry Ford, actor Mickey Rooney, and studio executive Samuel B. Mayer (of Metro Goldwyn Mayer) posed on a 4-person bicycle in the museum about 1942 (left).

In 1997, actor and race car owner **Paul Newman** was photographed with the 1906 racer known as *Old 16* (below).

Astronaut Neil Armstrong, first man to walk on the moon, is shown in front of the Wright Brothers' home about 1972.

Special events thrived during the 1960s and '70s. One of the most popular was the Old Car Festival, which began in the fall of 1951 with 88 participants. Visitors could see old cars moving and could talk to the owners about their machines.

A consultant's report commissioned in 1947, the year of Henry Ford's death, confirmed the success of his initial vision:

"The Edison Institute Museums (including the Museum Building and Greenfield Village) have probably the largest existing collection of authentic material illustrating the industrial and cultural development of the country, including manufacturing, agriculture, transportation, communication, and the applied arts....These museums can make an educational contribution surpassing that of any other institutions in the museum field."

But what happened immediately after Ford's death? Frankly, not very much. Henry Ford had never created an official chain of command or a management system beyond his own personal say-so. Assembled from the Ford Farms, the Trade School, and the automotive company, the staff developed their own personal methods of running their individual areas. The museum and village were operated almost as two separate organizations, neither running very efficiently or effectively. One manager recalled, "Mr. Ford said more than once he didn't want to make any money out of the project, so we weren't worried about that end of it." In fact, so unconcerned were Ford and his successors about expenses that each visitor during those early years *cost* the organization an average of $5 each.

Clara Ford took over for her deceased husband. As Hayward S. Ablewhite, who was appointed museum director in 1949, recalled: "No one else had anything to do with it.... But she would come over...at least once a week...and would be very disturbed if she thought you had any ideas of doing anything that she would think contrary to what Mr. Ford's ideas for the museum were."

THE TRANSITION

1947-1980

The Muzzle Loaders' Festival began as the Greenfield Village Turkey Shoot in November 1955. By 1975, the competitive shoot by Civil War re-enactors included the firing of cannons, which created the noise and smells of a nineteenth-century muster day. Safety concerns brought the event to a close in the early 1980s.

Two museum traveling exhibits followed the successful *Industrial Progress, U.S.A.* in the early 1950s. *Schoolroom Progress, U.S.A.*, focusing on educational advances since the days of the one-room school, hit the rails in 1955 after the official dedication of the two cars in Washington, D.C. It was followed by *Main Street, U.S.A.* and, like *Schoolroom Progress*, it traveled around the country for five years.

On October 25, 1955, the *Today* show beamed chapel services to viewers across the country. In mid-morning, mid-nineteenth century village life was portrayed on Arlene Francis' *Home* show. *The Howdy Doody Show* aired during the afternoon from the Scotch Settlement School.

Mrs. Ford had been the third founder and trustee of the organization, along with Henry and their son, Edsel. With her death in 1950, it truly was time for a management revolution.

The first formal organizational chart was drawn up in 1950. New executive positions were created and lines of responsibility and authority outlined. A.K. Mills served as executive director from 1951 until 1954, when Donald A. Shelley replaced him. Shelley served in that position until 1968, when his title was changed to president, a position he held until 1976.

Without Henry Ford's personal wealth and authority, the collection, already incredible in size and scope, was embellished by the addition of significant, often expensive, pieces of folk art, furniture and decorative arts. The college-level Edison Institute of Technology had been closed during World War II, and the high school in 1952, leaving only the elementary school in operation. The public, especially the vast travelling public of the 1950s, became the focus of attention.

A series of special public extravaganzas, more attentive to pageantry than accuracy, was initiated: "The Country Fair of Yesteryear" debuted in 1951 and was held annually for the next 30 years. The "Old Time Turkey Shoot" was soon renamed the "Muzzle Loaders' Festival." The "Old Car Festival" made its first appearance in 1951, and by the end of the century was the longest running antique car gathering in the world.

In 1953, the museum building was officially renamed Henry Ford Museum. That same year, the first major addition to the museum opened with an exhibit on the founder, *Henry Ford: A Personal History*. Special thematic exhibits on topics like early American music, sports cars and lighting devices offered new ways of looking at the collections. The Ford Motor Company donated the nearby Dearborn Inn, one of the nation's first airport hotels, to the museum as a source of income.

Buildings in Greenfield Village previously used exclusively by the school were opened to the public. Opening the Noah Webster House, Sarah Jordan's boarding house, and the Grimm Jewelry Store increased the variety of buildings available for public viewing. Improved craft demonstrations became an even more prominent feature of the visitor experience. The village became more inviting when the Clinton Inn cafeteria, previously used only by the school, was opened to the public. The all-white building exteriors that had dominated the landscape at Henry Ford's insistence were changed to pleasant, although historically inaccurate, colors.

For the first time, the institution also extended its public programs beyond the Dearborn area. In 1952, a 30-foot trailer truck carrying *Industrial Progress, U.S.A.* began a three-year national tour. *Schoolroom Progress, U.S.A.* and *Main Street, U.S.A.* followed as railroad car exhibits that each toured the country for

Village staff put the finishing touches on the re-constructed Heinz House. The building was originally from Sharpsburg, Pennsylvania, and as the home of H.J. Heinz was the first production site for his commercial food products. Later, in 1904, when Heinz was a famous name, the house was put on a barge and floated down the Allegheny River to be an exhibit at the H.J. Heinz Company Pittsburgh factory. In 1953 it was moved again, this time brick by brick, to Greenfield Village.

Every year, the Edison Institute School students put on elaborate plays in the museum theater. Although the high school graduated its last class in 1952, the elementary school continued in operation until 1969. This photo shows a scene from a holiday play around 1950.

Then & Now
Susquehanna Plantation House

Henry Ford wanted to include some aspects of the history of the American South in Greenfield Village, and was offered the Susquehanna Plantation house in 1942. It was originally located in St. Mary's County, in the tidewater region of Maryland near the Susquehanna River and the Chesapeake Bay. When Edward Cutler went to inspect the building, he found it intact but run down. Inside, he had to wade through 18 inches of grain to take his measurements. The house was restored in the village during the summer of 1942.

The design of the house, with its long, wide porches on either side and opposing windows and doors for flow-through ventilation, is typical of the tidewater region. When Ford acquired the building, he had been led to believe that the house was constructed around 1650. Recent historical and archaeological research has revealed that it was actually built around 1840. The house has been restored to its appearance in 1860 when it was the home of planter Henry J. Carroll, his wife and five children. Carroll owned the 700-acre plantation as well as the 75 enslaved African Americans who were required to keep it running.

As this image makes clear, rows and rows of everything were the typical display technique before the late 1970s. Henry Ford wanted to display all of his collections; therefore, he provided no storage space. The 1914 Detroit Electric, Clara Ford's personal car, is shown here next to a collection of license plates from every state of the union for the year 1929, the year the museum was dedicated.

several years. Befitting the political climate of the era, each exhibit extolled the idea of constant and universal progress in America.

By the 1960s the institution seemed to reach its stride. Expansion and experimentation marked the next decade of public programming. Extensive investments in producing television and school audiovisual programs illustrated the difficulties of translating the museum experience, up close to real objects, into purely visual media. The Edison Institute elementary school was closed in 1969, in favor of more attention to the general public and public school audiences. A growing staff of museum professionals regularly produced significant additions to the collections, exhibitions and special events. Attendance rose from 500,000 in 1950 to 1 million in 1960 and 1.3 million in 1969.

In 1964, the Ford Motor Company officially donated the millions of documents, papers and photographs comprising the Ford Archives. Containing both personal and business records, this is one of the most complete and significant business archives in the country. Curators in this era showed particular interest in folk and fine arts, adding hundreds of expensive, highly collectable items to the decorative arts and furniture collections. Few buildings were added to the village after Ford's death, but the Daggett home, an eighteenth-century Connecticut house of the "middling sort," was undoubtedly the most important.

In 1966 William Clay Ford, Henry's youngest grandson and then

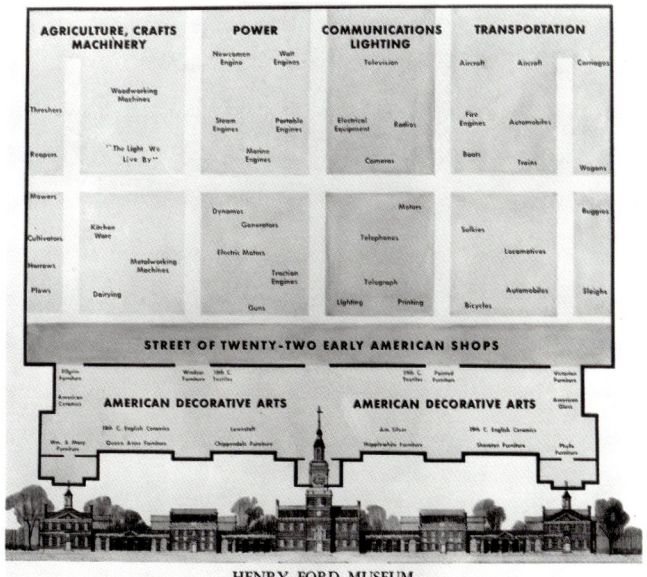

This map illustrates the organization of the collections and their layout in Henry Ford Museum as they appeared in 1962. Not much had changed since the 1940s and it would be another ten years before a major reorganization would take place.

president of the museum's Board of Trustees, oversaw the formal separation of The Edison Institute from the control of the automotive company that bears the family name. By 1969, the Ford Motor Company and The Ford Foundation had each made $20 million grants to the museum. In speaking at the institution's fortieth anniversary celebration that year, Ford said,

"I think the institute is one of the great philanthropic legacies of my grandfather. It in no way diminishes the significance of this historic resource to note that he underestimated its financial needs when he conceived it more than a generation ago."

For the first time, an endowment fund was created to provide for future income. Nearly half of the grant money was used immediately for necessary upgrades to utilities and improvements to the program facilities. The Great Hall of the museum was reorganized with color-coded carpet designating exhibit areas, uniform label styles, and the first major reduction in the number of objects on display. Henry Ford's philosophy of keeping everything on display was revised and the museum's first collection storage building finally built.

First, however, the institution had to recover from a fire that destroyed several of the shop displays and other exhibits in the museum in 1970. This fire was one of the most destructive ever to hit an American museum, destroying hundreds of irreplaceable artifacts, including major portions of the textiles and quilt collections. The fire brought to light a major museum shortcoming.

The 1970 fire of unknown origin caused extensive damage, including smoke damage to collection items far from the actual fire. While the museum re-opened in two days, it was a full year before all the exhibits were cleaned or re-built.

Typical of the 1970s reorganization of the museum, the lighting and communication exhibits used color-coded carpet and exhibit case interiors to display large, important collections (far left, top). For the first time, the cases displaying the lighting devices contained identification labels and general introductions, and were lighted.

One of five Focal Points that were part of the redevelopment of the museum in the early 1970s, the recreated kitchens in the Home Arts section (far left, below) continue to be popular with visitors.

Staff of the Crafts Department (left) demonstrated craft processes in the museum's Street of Shops as well as the Village Craft Area. Craft products were sold in the museum stores.

In 1956, the museum acquired the 600-ton Allegheny Locomotive built in Lima, Ohio, in 1941. It took 18 days to move the locomotive into the museum, and included laying new track, enlarging the museum's back doors, and taking parts off of the cab.

Never a stickler for paperwork, Henry Ford had not instituted a consistent system of record keeping for the collection. And the staff had made limited progress since his death. Curators could not even positively identify what was lost in the fire.

An optimistic business forecast projected an annual attendance of three million visitors. Further expansion seemed in order. A circular railroad was completed in 1972; a re-creation of a turn-of-the-century amusement park complete with a 1913 carousel, Suwanee Park, was completed in 1974, as was the re-organization of the working craft buildings.

Now, the organization had to fully fund itself through endowment income, contributions, and earned revenue. Thankfully, the Bicentennial Year of 1976 saw an all-time attendance high of 1.8 million visitors. Still, the energy-conscious 1970s were not the best time to rely on income from visitors to Detroit from around the nation. The cost of expanded public programming and the maintenance of the aging museum and nearly 100 village structures constantly exceeded revenue. The organization had vastly expanded its public popularity but at real cost in financial stability and educational depth. ■

As part of an effort to develop a membership audience for Henry Ford Museum & Greenfield Village, the institution took advantage of talented staff and the increased interest in enrichment classes by the public.

Museum President Donald Shelley and William Clay Ford, President of the Board of Trustees, shake hands to commemorate the completion of the village railroad in August 1972.

In 1951, the steamboat *Suwanee* (left) was overhauled and became the focus of new special programming around the Suwanee Lagoon and Activities Field.

The Suwanee Park project, finally completed in 1974 (below), included a bandstand, railroad station, ice cream parlor, penny arcade, carousel, gift shop and restaurant adjacent to the lagoon and the *Suwanee* steamboat.

The Bicentennial of the Revolution heightened Americans' interest in history and provided the opportunity for massive public programming in the museum and village. A record-setting 1.8 million people visited the museum and village.

Then & Now
Daggett Farmhouse

An entire eighteenth-century Connecticut "saltbox" style farmhouse and furnishings were donated by Mary Dana Wells to the village in 1977. The house had to be moved from Union, Connecticut, where its donor had relocated it from nearby Andover. As the Daggett Farmhouse, it now anchors one end of the village's agricultural offerings. Costumed staff tend the gardens, prepare meals, and spin yarn.

Famous Friends

Famous Friends

In 1992, civil rights pioneer and Detroit resident **Rosa Parks** (opposite) spoke with school students in the parlor of the Mattox House, the 1930s Georgia home of an African-American family.

Detroit's favorite son, heavyweight boxing champion **Joe Louis**, posed for a tintype (above) in the Greenfield Village Tintype Studio about 1940.

Former President **Gerald R. Ford** (right) joined Museum President Harold K. Skramstad, Jr. and costumed interpreters for the dedication of the Firestone Farm in 1985.

President **George Bush** is shown preparing to take off from the museum's parking lot in October 1992 (right). President Bush came to the museum to tape an interview for a television network news show.

"The Ford Museum is still a private organization marching to the beat of its own drummer, and a frustrating, woeful example of missed opportunity …[t]he collections still await a board of trustees, an administration, a curatorial staff and an exhibit philosophy and design that will do them justice."

Technology and Culture, 1980

Though no one was fully aware of it at the time, something quite extraordinary happened in the meetings of the museum's Board of Trustees in 1980. President Frank Caddy, who had begun his career in the village crafts department in 1931, decided to step down. The Board, encouraged by its youngest member, Sheila Ford, agreed to look far beyond the confines of Michigan, let alone the village, for a new type of leader.

In 1981, Harold K. Skramstad, Jr., then President of the Chicago Historical Society, agreed to come on board as the new president. Just 41, Skramstad brought a Ph.D. in American Civilization and vast experience at the Smithsonian's National Museum of American History as well as the Chicago Historical Society. Skramstad was steeped in the new professionalism of a national museum community. This new generation of professionals combined scholarly rigor with a passion for popular programming. They also developed a keen sense of the financial limits they were working under.

The museum he inherited showed potential but not excellence in those areas. As he himself said, he was drawn by the potentiality of the extraordinary collection of artifacts amassed over the years. But it was a collection that did not seem to add up to anything, nor did the programming help make sense of it for visitors. A 1978 article in *American Heritage*

The 1987 *Automobile in American Life* exhibit was the first dramatic change in the museum's design and thematic approach in more than a generation.

THE TRANSFORMATION

concluded: "Ford's reconstruction of the American past in Dearborn is still more than a little puzzling. Irony and paradox are everywhere. This is a record of something, but it is hard to say of what…an immense falsification of the historical past."

Skramstad began on an inauspicious note. The institution had been running an operating deficit for several years and its endowment was entirely invested in Ford Motor Company stock. In 1982, the company, rocked by shifts in the global economy, declared that there would be no dividend. This removed another $2.5 million from the budget. Skramstad later noted that the first question he faced was whether the institution would survive or not. He trimmed the staff by 30 percent, doubled the admission fee and instituted new financial controls. He sold the Dearborn Inn to increase the endowment. He diversified the endowment portfolio and also found new sources of financial support, successfully tapping in to the National Endowment for the Humanities and private foundations in ways unprecedented for the museum.

Other changes were afoot as well. Members of the Ford family, several of them in the fourth generation, showed renewed support for the organization and faith in its potential for civic influence. Simultaneously, they recognized that a complex enterprise with expanding community ties was best served by a Board of Trustees that included local business and civic leaders as well as family members. A staff curriculum committee surveyed the collections and programs in order to provide a new

Harold K. Skramstad, Jr., president from 1981 to 1996, brought energy, professionalism, and imaginative ideas to the museum. Staff often joked that the four scariest words in the world were, "Harold has an idea." For his work at the museum as well as his contributions to promoting the humanities in public life, he was honored in 1992 with the prestigious Charles Frankel Award from the National Endowment for the Humanities.

By the 1980s, the village's flagship installation, Thomas Edison's Menlo Park laboratory complex, was showing the wear from 60 years of public visitation. Curatorial and conservation staff cleaned and conserved irreplacable Edison artifacts. They renovated the buildings, including the business office shown here, opening them up more fully and furnishing them more completely than they had been.

Streamlining America was a popular temporary exhibit in the museum in 1986 and 1987. It dealt with the way the streamlined design became a widely used symbol of modern America consumerism in the 1920s and 1930s. This exhibit began the museum's serious attention to more recent twentieth-century history.

Henry Ford had shown an early interest in presenting the history of African Americans in Greenfield Village through his inclusion of the George Washington Carver Memorial, two slave houses, and the Mattox House, the Georgia farmhouse owned by generations of the Mattox family. In 1991, each of these buildings was renovated and new programs on African-American Family Life and Culture initiated.

focus and direction to the public experience. The committee concluded that the institution could most powerfully tell the story of America's transition from a rural, agricultural society to an urban, industrial one. Concentrating on the colonial past may have brought attention during the Bicentennial of the American Revolution-fixated 1970s, but a colonial focus was not making the best use of the collections.

Based on this report, curators began reviewing the collections to see how well they told the story of change in America. What should be added? What objects were irrelevant or unusable? While adding to the collections has never stopped, the emphasis has been on refining the scale and scope of collections and making room for new acquisitions.

Since collection items are held in public trust, their disposal must meet strict legal and ethical standards common to American museums. Thousands of objects, even entire buildings, have been removed as a result of this process, which is continuing into the twenty-first century. They are sold at public auction or placed with another nonprofit organization. The resulting funds are invested in an endowment account that supports the acquisition and care of collection items.

A computerized collections management system was installed in the late 1980s to help organize and provide access to the millions of pieces of information relating to collection items. Storage areas and conservation efforts received substantial resources and staff attention. Still, the simple

physical needs of this huge, unique collection of historic objects continued to strain the organization's financial resources.

Small exhibits in the museum focused on historical themes, such as the design of mass-produced consumer goods in the late nineteenth century. Buildings in the village began to change appearance and presentations as curators trained in social history furnished them "as lived in" rather than as galleries of decorative arts.

Changes were made in the village in 1984 when existing buildings were moved to make way for the Firestone Farm. This working seven-acre farmstead included a nineteenth-century house and barn moved brick by brick and board by board from Columbiana County, Ohio. Visitors were greeted by a view of the farm fields with sheep and cattle when first entering the village. They would learn of the fundamental work and lifestyle that sustained 80 percent of all Americans up to the twentieth century from costumed presenters who actually raised the crops, fed the animals, cooked the meals, and cleaned the kitchen and barn.

The drive toward historical authenticity reached venerable installations as well as the new ones. In 1987, curators removed the display cases containing a tree stump and twentieth-century refuse unearthed by Henry Ford at Thomas Edison's laboratory at Menlo Park, New Jersey. Dozens of framed photos of Edison and Ford were taken down from the walls of the laboratory building. The site was re-interpreted to look as it did in 1880, when Edison was completing development of the light bulb system, and the story focused on Edison's invention process.

However, the most visible coming

The Automobile in American Life (1987) was a revolutionary renovation of a main museum exhibition. Examining the automobile industry's evolution and impact, it was organized into five distinct thematic areas with uniquely designed floor plans, lighting and graphics.

In 1992, the museum embodied its new mission emphasis on innovation with a giant participatory activity center, "The Innovation Station."

Then & Now
Firestone Farm

A significant addition to the historic structures in Greenfield Village was the Firestone Farmstead from Columbiana County, Ohio. Built in 1828, the farmhouse was the birthplace of Harvey Firestone, the founder of the Firestone Tire and Rubber Company, and one of Ford's lifelong friends. The Firestone farmhouse (shown here as it looked in the late 1870s) was dismantled brick by brick in 1984 and moved along with a large barn to Greenfield Village.

Horsepower, farm stock and crops, family work, and seasonal change are staples of the visitor experience at Firestone Farm.

of age of the new generation of public presentations was *The Automobile in American Life*, a 60,000-square-foot exhibition in the museum. Where previous displays had presented objects of just one type, this new design "look" combined cars, car parts, paper documents and books, photographs and banners, and media presentations in interpretive settings. One setting even evoked an outdoor camping scene. A real 10-foot by 15-foot tourist cabin from the 1930s was juxtaposed with a room from a 1950s Holiday Inn, and a full-sized 1940s diner was placed next to a 20-foot tall McDonald's sign.

Initially unpopular with car collectors, largely because the number of automobiles on exhibit was reduced from 220 to 110, the exhibit proved popular with the general public and with professional historians. Attendance increased in 1988, and a prestigious exhibit award from the Society for the History of Technology provided scholarly respect.

The renovation of the power and machinery collections into the 50,000-square-foot *Made in America* exhibit followed a similar style and educational approach. Objects became part of a historical story made as familiar as possible to everyone. Between them, the automobile and industrial history exhibits were the two largest exhibits on a single topic in the country and larger than most other museums.

The Innovation Station, a tri-level activity center, employed a totally different approach by making visitors an active part of the experience itself. Rather than display historic artifacts, the Innovation Station was a response to public demand for involvement

A 1995 partnership between the museum and the Motown Historical Museum produced *The Motown Sound: The Music and the Story* and the restoration of Hitsville, USA, in Detroit. Visitors engaged in activities—mixed a recording, played disc jockey, learned to dance like The Temptations—as well as learned of the extraordinary vision and drive of Motown founder Berry Gordy, Jr.

In 1992, *Made in America* opened as the world's largest and most comprehensive exhibition of American industrial development. From the world's oldest steam engine to modern robotics, the exhibit illustrated the transformation of the ways we make things.

A new exhibit on the life and accomplishments of Henry Ford was installed in the center of the museum in 1996. Entitled *Henry's Story: The Making of an Innovator*, the centerpiece of the display is the 1896 Quadricycle, the first car Henry Ford built.

and participation. By the end of 1992, over 115,000 square feet of the museum (about one-third of its exhibit space) and nearly half of the buildings in Greenfield Village had been researched, re-furnished and "re-furbished" with new presentations.

Underlying this transformation of the public experience were changes in the museum's mission and the way the staff was organized and worked. In 1992, Skramstad invoked another look at the museum's meaning and messages. Responding to changes in the marketplace of leisure time and ideas, the trustees adopted a new mission statement: "Henry Ford Museum & Greenfield Village provides unique educational experiences based on authentic objects, stories, and lives from America's traditions of ingenuity, resourcefulness and innovation. Our purpose is to inspire people to learn from these traditions to help shape a better future."

In 1993, a transition team headed by Steven K. Hamp recommended massive changes in the organization's structure and work processes. The result was the formation of business units that had direct contact with customers and responsibility for revenue generation alongside support units that provided essential internal services. Staff training focused on teamwork and communication rather than traditional management control. The new managerial and work methods were a response to new financial and public demands for renewed experiences.

In 1985, only the Finance Department had a computer, an

Putting a new roof, doors, and windows on the museum building was a herculean task, taking more than three years and $7 million. The addition of a new heating and air conditioning system cost another $8 million, involved over a mile of new duct work, and required mobile home-sized chilling units to be hoisted to the roof using the nation's largest mobile crane.

Called the "Mona Lisa of American automobiles," *Old 16* was the first American car to win a major international race. Winner of the Vanderbilt Cup in 1908, the Locomobile race car put the infant American automobile industry on the world map. Obtained from a private collector in 1997, the legendary car went on long-term public display for the first time.

unwieldy mainframe cast-off. By 1994, the museum had made a commitment to computerization of its business practices and invested a million dollars in desktop computers, networking and training for more than 150 staff members. Major grants enabled the development of a multi-media computer lab and distance learning theater for educational programs. By 1998, the organization had a five-person unit dedicated to maintaining and servicing these electronic communication systems. The new computerized telephone "switchboard" was placed on public display as a symbol of the museum's modernization.

In the summer of 1996, Skramstad decided to step down as president. In his 16 years at the helm, the institution had moved in remarkable ways. Not everyone agreed with all the changes, of course. A 1987 news magazine headline had screamed: "Under Fire. Harold Skramstad: Is He Betraying or Realizing Henry Ford's Vision?" Yet, as Skramstad was leaving, most people acknowledged that he had tapped into the institution's immense potential and left it with even more opportunities than when he arrived.

The institute's financial situation was strong (helped in no small part by a diversified and growing endowment, supported by a strong domestic economy in the 1990s), attendance had leveled off at just over one million visits a year, and the institution's local and national reputation had never been stronger. And the organization was on the verge of opening an on-site public school, Henry Ford Academy. Skramstad's tenure had

The 1913 Herschell-Spillman carousel, a favorite of village visitors for over a quarter of a century, found a new home near the Village Green in 1998. The new structure, especially designed for the carousel, provided improved care for the elegant but fragile painted animals and Wurlitzer band organ as well as amenities like food service for visitors.

concentrated on programming for broad audiences, high fiscal and intellectual standards, and a sense that museums were about using the past in the present.

Steven K. Hamp succeeded Skramstad in one of the fastest, least disruptive transitions in the history of major museums. The museum's Board of Trustees accepted Skramstad's resignation and one hour later named Hamp the new president. Hamp and others saw in his appointment a ratification of the directions the museum had been taking under Skramstad. But in at least one way, his appointment was a throwback, for Hamp had spent most of his professional career at the museum, leading the archives, collections, and educational programs units since joining the staff in 1979.

Now, Hamp had to turn the organization's and the public's attention even further behind the scenes. The largest capital campaign in the organization's history was initiated. It aimed to improve the ancient infrastructure that Henry Ford had constructed in the 1920s and 1930s. The museum roof and 90 exterior windows were replaced, and the original heating system of hot-water radiators was replaced by more than a mile of overhead duct work running the length of the museum, capable of providing air conditioning, humidity and air quality controls. The air-conditioned interior provides a distinct benefit for collections and visitors alike.

Changes to the village included moving the boat dock and carousel up to areas adjacent to the Village Green. A revised railroad system included a new station, connecting to the adjacent Amtrak system. Bringing modern communication systems to

The annual July 4th "Salute to America" weekend concerts by the Detroit Symphony Orchestra at Greenfield Village became a public phenomenon, and a potent partnership between two of the area's major cultural organizations.

Then & Now
The Lincoln Chair

One of the most compelling artifacts in the museum's collections is the Lincoln Chair—the chair President Abraham Lincoln was sitting in when he was assassinated on April 15, 1865. This photograph (near right) was taken shortly after the murder by famed Civil War photographer Mathew B. Brady.

The chair originally belonged to Harry Clay Ford (no relation to Henry), whose brother managed Ford's Theater in Washington, D.C. Harry brought the chair from his own home so the president would be more comfortable while watching the play. After the assassination, all of the furniture used by the presidential party was seized by the government as evidence.

The chair remained in government storage until the late 1920s. Then, Harry Ford's wife went to court to argue that the government never had clear title to the chair. Eventually, it was returned to her. The chair was put up for sale and Henry Ford purchased it in 1929. It was shipped to Dearborn in a large crate and unpacked in the Logan County Courthouse (above, right); the chair was displayed there until the late 1970s, when it was moved to the museum.

In 1999, the chair was thoroughly examined and its fragile upholstery cleaned and conserved. Then the chair was given a secure, climate-controlled display case.

village sites, providing climate control to important structures, and building facilities for the Academy were central elements of the plan. Several new exhibitions, both temporary and long-term, were elements in the $40 million campaign, "Shaping the Future While Preserving the Past."

Opened in 1997, the Henry Ford Academy, the public charter high school, was an important feature of this reinvigorated museum. Hamp conceived the school as a return to the founder's initial vision. He also saw it as a demonstration of the museum's modern commitment to educational innovation and community improvement. Co-sponsored by the Ford Motor Company and chartered by the local public school system through the Wayne County Regional Educational Service Agency, the school occupied space in the museum and village, making it among the most publicly visible schools in the nation.

The 400 students in grades 9-12 come from throughout Wayne County and represent the diversity of the area. They learn in equally diverse ways, making use of the museum's resources as well as materials available electronically on the World Wide Web and the Ford Motor Company staff who mentor students. Student work is being invented and reinvented

The Mattox Family Home, originally from rural Bryan County, Georgia, southwest of Savannah, is restored to its 1930 appearance. Initially thought to be the house of a white overseer, research in the 1980s revealed that it had been built by a land-owning African-American farmer in the late 1800s. Mattox family photos and furnishings remain in the house, while staff raise chickens, tend the garden, and cook in the kitchen.

Mike Vliet, in cap and gown, posed at Town Hall for his 1997 graduation picture with program leaders and fellow students in the nationally acclaimed Youth Mentorship Program. Vliet was the first mentorship student to earn all of his elective credits through the museum program and the first to go on to college.

constantly. The goal is to engage students in rigorous academic work that enables a lifelong commitment to learning, resourcefulness and innovation in problem solving, and respect for themselves, their communities, and the environment. The innovative collaboration between the museum, the company and the public schools drew national attention.

The institution regularly ranked at the top of Michigan's cultural organizations in program quality and public attendance. It was recognized as a leader among American museums by the American Association of Museums. In 1998, the museum's youth mentorship program earned the institution the National Museum Service Award. In the same year, *The Detroit News* declared that Henry Ford Museum & Greenfield Village was the only local museum worthy of "world-class" status.

The 1999 construction and opening of a giant-screen IMAX® Theatre provided a major new venue for presenting America's innovation and ingenuity, as well as a strikingly different visitor experience. Combining with nearby separate attractions, the Automobile Hall of Fame and the Spirit of Ford, the museum's IMAX Theatre contributed to the creation of a new "cultural campus" anchored by the existing museum and village.

The last two decades of the twentieth century have proven the durability of Henry Ford's instinct for collecting the commonplace and his initial vision of an innovative educational institution. Since his time, each generation has done its best to give contemporary power and dimension to that vision. That opportunity will continue into the twenty-first century. ■

The most significant structural change since 1929 to the museum, itself a National Historic Landmark, will be the IMAX Theatre and Visitor Reception Area opening in late 1999. The giant-screen theater with a 62- by 80-foot screen capable of showing two- or three-dimensional films, wrap-around digital sound system, and seating for 400 offers a new, sensory-rich experience for visitors.

When Thomas Edison perfected a commercially successful phonograph 11 years after first inventing the hand-cranked version in 1877, he was reinventing his own invention. He was also creating a new industry, the musical recording industry. Henry Ford did the same thing for the automobile when he developed the moving assembly line at his Highland Park plant in 1913, thus reinventing the automobile industry. He then reinvented the industry again in the 1920s with the integrated, raw-materials-to-finished-car system at the massive Ford Rouge complex. More than mere process improvements or even new product lines, reinventions expand opportunity by transforming something well-known into something absolutely original, innovative and influential.

Americans in the 1920s and 1930s were amazed by the audacious focus and scale of Henry Ford's historical enterprise. They were enthralled by the variety of material goods their ancestors had used, and delighted in the environment Ford created in Greenfield Village. Ford's new kind of museum was unparalleled and unrivaled. Families have never tired of the magic of a child's first ride on the carousel, the astonishment at hearing the crude Edison tinfoil phonograph, or the gasp of amazement at the sheer size of the Allegheny locomotive. While educational fads come and go, students and teachers remain intrigued, even more today than a generation ago, by the educational and inspirational power of the museum environment.

Reinventing Henry Ford Museum & Greenfield Village for the twenty-first century will require the same type of vision, risk-taking, ingenuity and resourcefulness demonstrated by people like Edison and Ford. Providing climate control for the benefit of

Students at the Henry Ford Academy prepare for their futures in college or the workplace by participating in a community of learners in a resource-rich environment.

THE REINVENTION

2000...

Inventor/philosopher Buckminster Fuller's Dymaxion House, an aluminum concept house from the post–World War II era, will be opened to the public in the museum in 2001.

visitors and collections alike after three generations without that amenity requires ingenuity. Rebuilding an 1880s railroad roundhouse into a working railroad program venue requires resourcefulness. It takes vision to invest in a state-of-the-art movie theatre, thus expanding the visitors' experiential opportunities. However, these changes will not reinvent the museum, for change is expected. In order to succeed, we must surprise ourselves as well as our visitors and supporters.

Reinventing the museum will demand new, and as yet unimagined, connections between museum experiences and visitor interests. Any true reinvention must create a different dynamic between museum programs and community needs. This reinvention will respond to the opportunities of digital technologies, the competitive demands on visitors' personal time, and the challenges of renewing our communities at the turn of the millennium. Most important, our reinvention will rely on the power of the past to inspire us toward a better future. ∎

The six-stall 1884 Detroit, Toledo & Milwaukee Roundhouse from Marshall, Michigan, will become the new home for the American Railways Program in Greenfield Village in 2000.

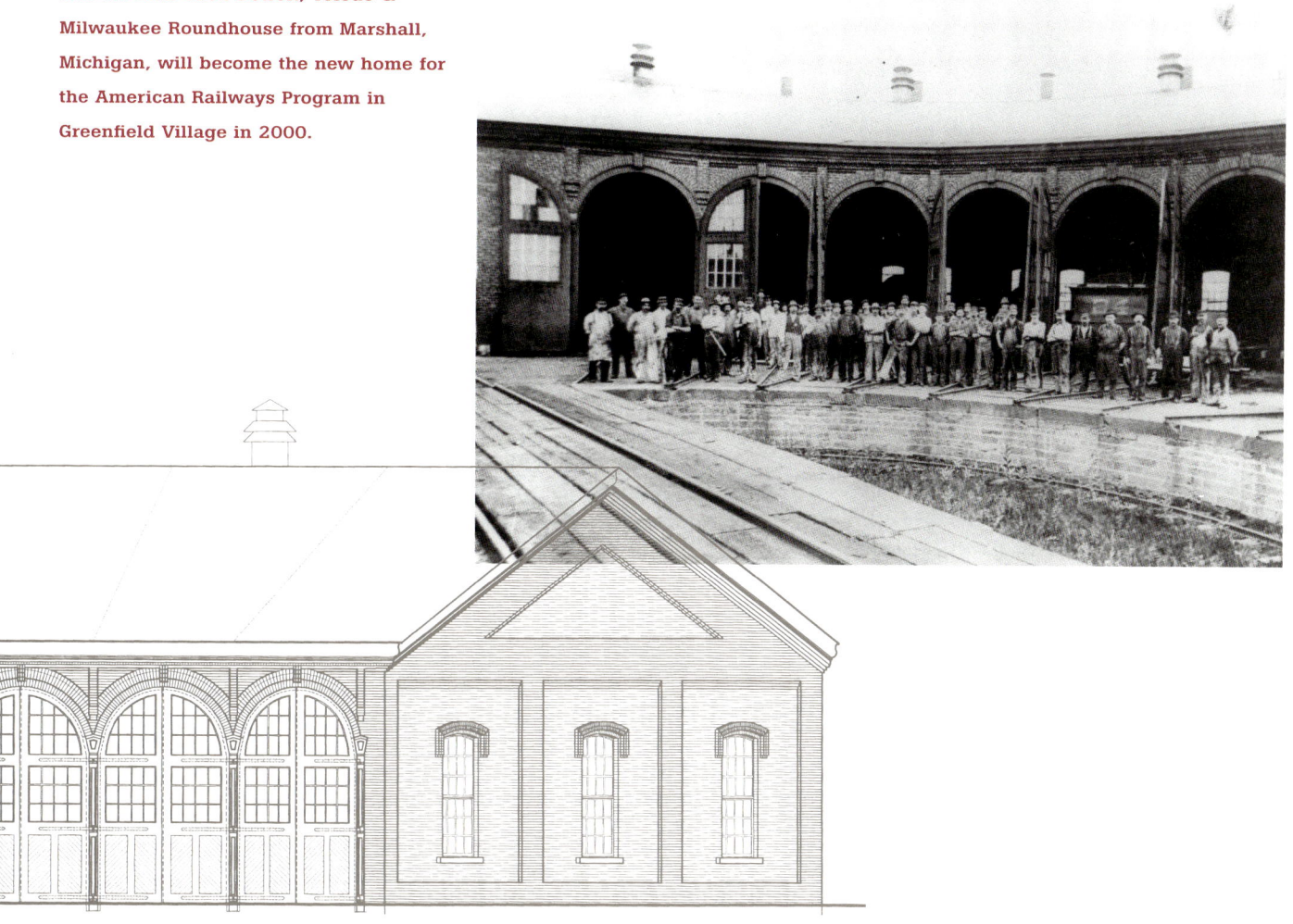

"One of America's strengths is its heritage. To neglect that heritage is to risk a future in which our young people in particular will find themselves without a means of building on the firm and reassuring foundation of the past."

William Clay Ford